YOUR KNOWLEDGE HAS VALUE

- We will publish your bachelor's and
  master's thesis, essays and papers

- Your own eBook and book -
  sold worldwide in all relevant shops

- Earn money with each sale

Upload your text at www.GRIN.com
and publish for free

Ataliba Miguel

Petroleum Geology Offshore Cabinda

An Overview of N'SANO Field

GRIN Verlag

Bibliografische Information der Deutschen Nationalbibliothek:

Die Deutsche Bibliothek verzeichnet diese Publikation in der Deutschen National-
bibliografie; detaillierte bibliografische Daten sind im Internet über http://dnb.d-
nb.de/ abrufbar.

Imprint:

Copyright © 2012 GRIN Verlag GmbH
Druck und Bindung: Books on Demand GmbH, Norderstedt Germany
ISBN: 978-3-656-62305-2

This book at GRIN:

http://www.grin.com/en/e-book/270708/petroleum-geology-offshore-cabinda

Petroleum System Offshore Cabinda
An Overview of N'SANO Field

Subsurface Report

Ataliba Miguel
13/04/2012
Word Count 2230

Summary

This report has been written to describe the basic understanding of the petroleum system on the N'SANO field offshore Cabinda in Angola. The N'SANO field was discovered in 1992 and is located on Block 0 offshore of Cabinda province in Angola. The Block 0 is mainly divided into three source rock Megasystems which are associated with three distinct geologic packages. The Bucomazi formation is the most prolific source rock in the Area A of Block 0. The source rocks for the Bucomazi are kerogen of type I and type I-II, the migration is mostly fault related (vertical), the reservoirs are sandstones of synrift and postrift age, sealed by transgressive marine shales of Cenomanian-Eocene age.

Key words: Petroleum geology; source rock; reservoir rock; migration; trap; seal

Ataliba Miguel

List of Contents

Abbreviations

1.	Introduction	4
1.1	Objectives	4
2.	Regional Geology	5
2.1	Stratigraphy	5
3.	Petroleum System - N'SANO Field	8
3.1	Source Rock	12
3.2	Migration	12
3.3	Reservoir rock	13
3.3.1	Upper Pinda Reservoir	13
3.3.2	Vermelha Reservoir	14
3.3.3	Likouala Reseroir	14
3.4	Seal	14
3.5	Trap	15
4.	Prospectivity Impact on Block 0 - NSANO Field	16
5.	Conclusion	17
6.	References	19

LIST OF FIGURES

Figure 1 – Location map of offshore Cabinda Block 0	4
Figure 2 – Generalized stratigraphic column for offshore Cabinda	7
Figure 3 – Main source rocks types and their origin	8
Figure 4 – Maturation process of organic matter	9
Figure 5 – Location map of Block 0 – N'SANO Field	10
Figure 6 – Geochemical log Bucomazi Megasystem	11
Figure 7 – Burial history	11
Figure 8 – Primary and Secondary migration.	13
Figure 9 – Structural trap	15
Figure 10 – Stratigraphic traps	15
Figure 11 – Anticlinal trap	16
Figure 12 – Events chart for the Congo Delta Composite	18
Figure 13 – General conceptual model of a Petroleum System	18

List of Tables

Table 1 – Major tectonic phases and stratigraphic events	6
Table 2 – Block 0 N'SANO field petroleum system description	17

Ataliba Miguel

1.   Introduction

The N'SANO field discovered in October 1992, is one the 21 fields
owned by Chevron in Block 0 located offshore Cabinda in Angola.
The Block 0 (Fig 1) has an extension from the coastline to about
200 m isobath and it is divided into three areas known as A, B, and
C (Schoellkopf and Patterson 2000, p. 361). The N'SANO field
located in offshore Cabinda province is at 75 meters of water in
Area A of Block 0 (King et al., 1998, p. 317).

The generation of hydrocarbons in Block 0 comes from multiple
sources. Three prolific source rock Megasystem have been
identified, which are associated with three geologic packages,
these are the lacustrine shales of Lower Cretaceous Bucomazi, the
marine Upper Cretaceous labe and Lower Tertiary Landana shales
and marls, and the Tertiary Malembo shales (Schoellkopf and
Patterson 2000, p.362).

1.1   Objectives

The objective of this report is to develop an understanding of the
petroleum system offshore Cabinda, specifically on N'SANO located
in Block 0 Area A.

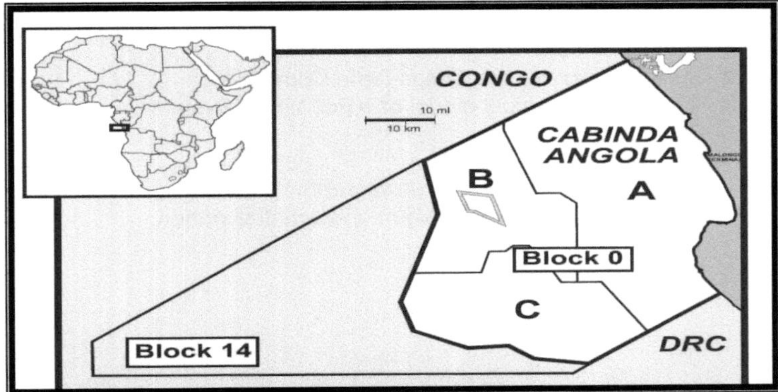

Figure 1— Location map of offshore Cabinda showing areas A, B,
and C of Block 0 (Schoellkopf and Patterson 2000).

                                                            Ataliba Miguel

2   Regional Geology

The offshore Angolan basins evolved from an active continental rift system during early cretaceous into a passive margin that culminated in the drift stage, which separated African from the South American plates (Machado, 2007).

The tectonic regime associated with the continental separation has been divided into four stages (Table 1), from oldest to youngest, (Cole et al., 2000, cited by Brice et al., 1982, p.326).

Firstly, a Prerift sequence occurred with the deposition of fluvial-lacustrine sands that is up to 1000 m in thickness (Cole et al., 2000, p.326). Secondly the Synrift stage characterized by graben development along the early rifted margin that formed the depocenters for organic-rich lacustrine shales. The lake systems created within the grabens were filled by lacustrine turbidites arranged laterally and upward into organic rich-shales; these are the shales of the Bucomazi Formation, the most active source rock for major oil accumulations in West African margin (Cole et al., 2000 cited by Burwood et al., 1995 and Burwood, 1999, p. 326). Thirdly, the Synrift II stage a period of regional subsidence has resulted in marine incursions, consisting of lacustrine carbonates and sandstones and alluvial clastics. At the end of this stage occurred the transitional sequence of the Aptian Loeme Salt formation, characterized by a widespread deposition of Evaporites. Finally the Post rift stage is characterized by an initial Albo-Cenomanian transgressive sequence as result from the thermal subsidence of the basins as the American and African plates separated.

2.1   Stratigraphy

Three distinct geologic packages are directly related with the stratigraphic section of Cabinda (Fig 2).

1) The Presalt nonmarine rift and sag section, 2) The salt and Postsalt marine Cretaceous–lower Tertiary drift and passive margin subsidence section, and 3) The Tertiary (Oligocene–Holocene)

Ataliba Miguel

sedimentary wedge (Schoellkopf and Patterson 2000, p. 362).

The Neocomian to Barremian Bucomazi formation is divided into three parts; 1. The Lower Bucomazi consists of laminated shale, mudstones, calcareous shale, marl, and sandstone with thickness up to 1000 m, 2. The Middle Bucomazi consists of organic-rich shale, carbonate mudstone, and marl with thicknesses ranging up to 1000 m, and 3. The Upper Bucomazi consists of green to brown mudstone and gray shale with thickness up to 400 m (Brownfield and Charpentier 2006, p.20)

Geological time		Age/Ma	Main tectonic stratigraphic events		Main depositional systems		Tectonic phase	
					Seaward	Shoreward		
Quaternary	Pleistocene	1.5	Drop in sea level/ regression / tilting Westward		Continental		IV	Post-rift (drift) stage
Neogene	Pliocene	5						
Neogene	Miocene	24	Rise in sea level/transgression		Marine			
Paleogene	Oligocene	37	Large drop in sea level / erosion/ tilting Westward		Marine	Continental turbidites		
Paleogene	Eocene	58						
Paleogene	Paleocene	65	Drop in sea level/regression		Continental sand wedges			
Cretaceous	Maastrichtian	74	Rise in sea level/transgression		Deep euxinic marine shales			
Cretaceous	Campanian	84						
Cretaceous	Santonian	87	Rise in sea level		Marine	Continental sand wedges		
Cretaceous	Coniacian	89						
Cretaceous	Turonian	91						
Cretaceous	Cenomanian	97	Walvis Ridge breached permanently		Shallow marine carbonate	Continental red beds		
Cretaceous	Albian	113	Subsidence of continent		Shallow marine carbonate			
Cretaceous	Aptian	119	Intermittent transgression		Evaporate/sands		III	Transition stage
Cretaceous	Barremian	124	Uplift/erosion				II	Syn-rift stage
Neocomian	Hauterivian	131	Deep euxinic lakes in rift		Fluvio-deltas/organic-rich shales			
Neocomian	Valanginian	138	Early rifting/Volcanism/Erosion		Lacustrine turbidites			
Neocomian	Berriasian	144	Doming/Flood basalts		Shallow continental basins		I	Pre-rift stage
Jurassic			Doming					

Table 1 — Major tectonic phases, stratigraphic events, and depositional stratigraphy in the west African — represented by Lower Congo Basin (LIU et al., 2008)

Ataliba Miguel

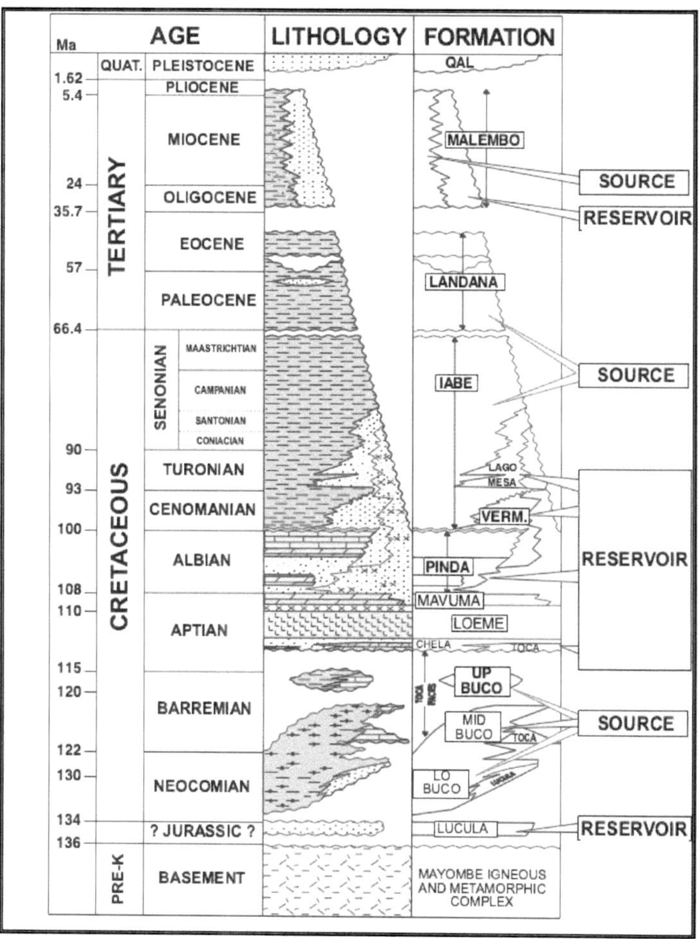

Figure 2 – Generalized stratigraphic column for offshore Cabinda.
(Schoellkopf and Patterson 2000)

Ataliba Miguel

3. Petroleum System – N'SANO Field

3.1 Source Rock

Source rocks are rocks containing organic material, mainly derived from Algae, Bacterial and Land plant. Potential source rocks are normally evaluated by the high concentration in organic matter, which is expressed as a percentage of TOC (Total Organic Carbon) value.

There are three known types of source rocks (Fig 3), 1. Kerogen type I (mainly algae derived), 2. Kerogen type II (mixed algae and bacteria), and 3. Kerogen type III (mainly landplant).

Sedimentary rocks formed from sediments rich in organic matter, once buried as a result of both sedimentation and subsidence, are exposed to increased temperature and pressure envelope. As a result of increased temperature and pressure, a biogenic decay aided by a bacteria and abiogenic reactions (in the organic matter) takes place in the shallow subsurface areas. Methane, carbon dioxide and water are given off, leaving the kerogen. When kerogen reaches the deeper subsurface, temperature and pressure are further increased, transforming the kerogen into petroleum (Fig 4).

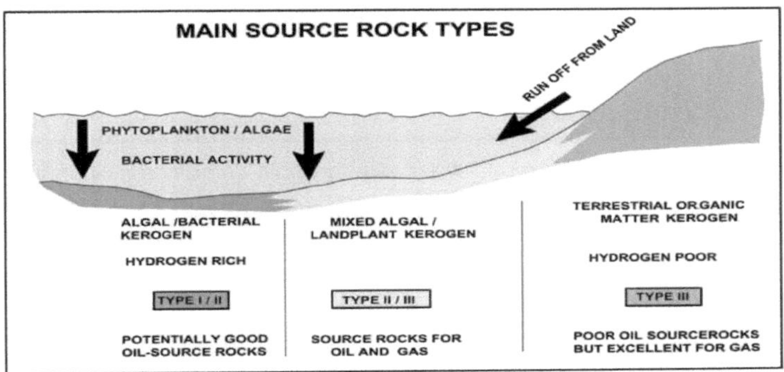

Figure 3 – Main source rocks types and their origin. (Vekeen 2007)

Ataliba Miguel

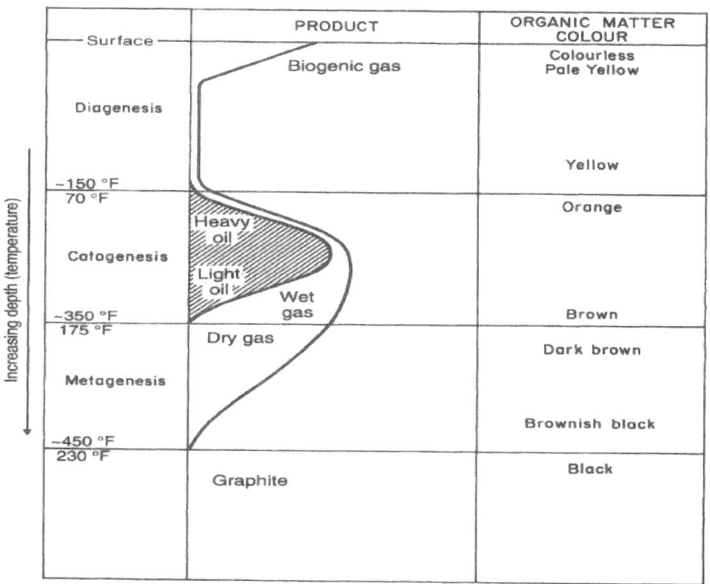

Figure 4 – Maturation process of organic matter (adapted from Stoneley 1995).

The Bucomazi formation is responsible for large quantities of oil found in Block 0 (Schoellkopf, and Patterson, 2000), specifically in the Area A where the N'SANO field is located (Fig 5).

Geochemical log from the Presalt interval (Fig 6) shows that the lower, middle and upper Bucomazi formation contains very good source rocks with a total organic carbon TOC >2%, and excellent hydrocarbon source potential $S_2>5$; the Kerogen is of the type I and type I-II, with high HI (hydrogen index) values (>500) and low oxygen indices (<100) (Schoellkopf and Patterson 2000, p. 364).

The maturity levels across Block 0 occurs in the Lower Bucomazi (Fig 6), where the values for the vitrinite reflactance and pyrolysis maturity are of $R_0 > 0.6\%$ and $T_{max} > 440C$ respectively, at a depth of 12,500 feet (3,750 m).

Ataliba Miguel

The timing of oil generation in Block 0 has been active from late cretaceous to the present (Schoellkopf and Patterson 2000, p. 374). For illustration purposes, the burial history graph was adapted from the offshore Congo formations as shown on Fig 7.

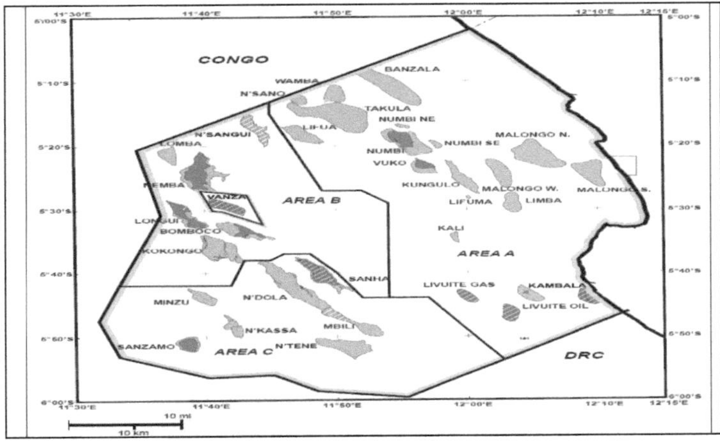

Figure 5— Location map of Block 0 – N'SANO Field (Schoellkopf and Patterson 2000)

Ataliba Miguel

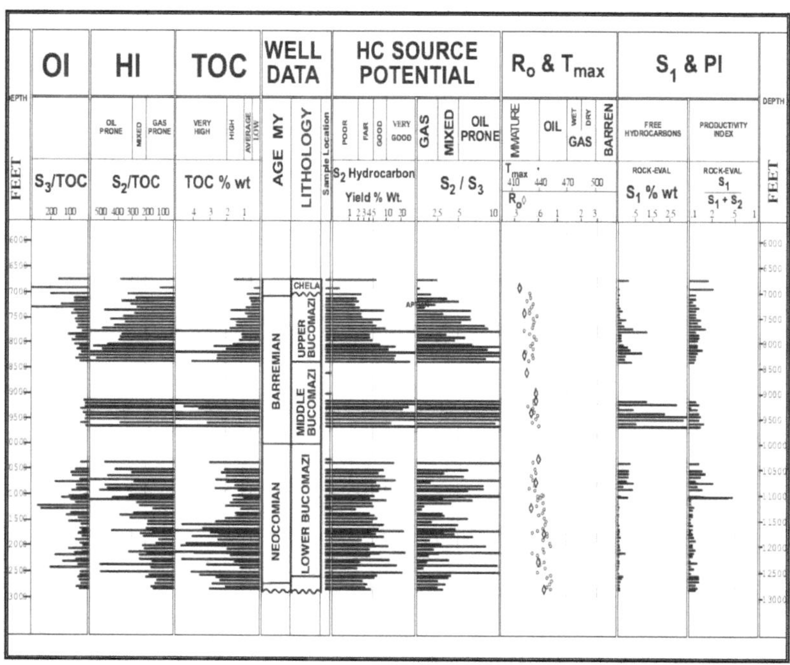

Figure 6 – Geochemical log from a selected well in the Lower Cretaceous Bucomazi Megasystem showing a typical signature for the Presalt interval (Schoellkopf and Patterson 2000)

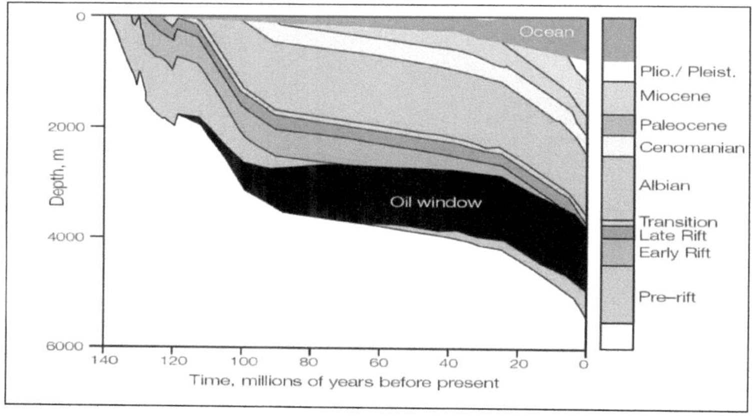

Figure 7 – Burial history (Schlumberger, 1993)

Ataliba Miguel

3.2   Migration

After kerogen has been transformed into hydrocarbon in the source rock, a significant amount of hydrocarbon is still maintained within the source rock; if the source rock is buried even deeper, the temperature and pressure will increase; hence the generated hydrocarbons are expelled off from the source rock and migrate away into more permeable strata (Fig 8). Normally the expulsion takes place at vitrinite reflactance R_0 of 0.6%. Once the hydrocarbons leave the source rock, they move freely along bedding planes until reaching a reservoir with good porosity and permeability characteristics. This is the onset of the migration of the hydrocarbons; migration path is determined by permeable layers and conduits in an up dip direction (Vekeen 2007, p. 284).

Schoellkopf and Patterson (2000, p. 374) concluded that although vertical (fault-related) migration pathways predominate in offshore Cabinda, significant lateral migration has occurred within the Presalt along the Chela formation. The availability of salt windows was very vital in the migration history of Bucomazi sourced oils; it provides migration pathways into the Postsalt interval. The salt windows occur in association with listric normal faults.

Ataliba Miguel

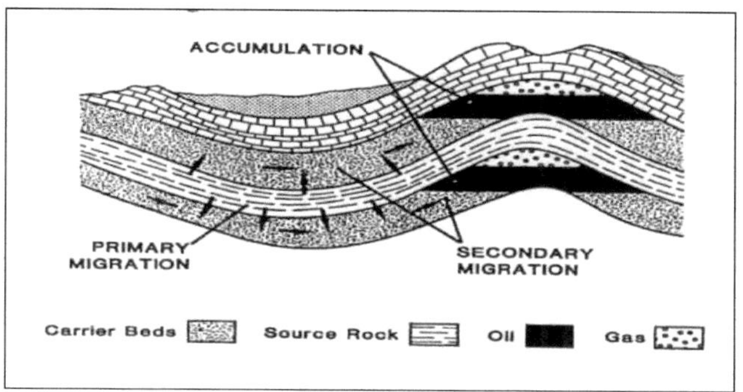

Figure 8 – Primary and Secondary migration. (adapted from England, Mann and Mann 1991)

3.3 Reservoir rock

Reservoir rocks are primarily classified as Sandstones, Limestone Dolomites and Chalk. Reservoir rocks in the shallower areas of the Congo Basin are mainly sandstones of both synrift and postrift age (Brownfield and Charpentier 2006, p. 39) that were deposited as eolian, alluvial and lacustrine turbidite sands and as lacustrine carbonate shoals (Schoellkopf and Patterson 2000, p. 368).

The reservoir structure of the N'SANO field comprises three non-communicating hydrocarbon-bearing reservoirs 1) Upper Pinda reservoir, 2) Vermelha Reservoir, and 3) Likouala Reservoir (King et al., 1998, p. 317).

3.3.1 Upper Pinda Reservoir

Pinda reservoir was deposited in a flat cyclical near-shore marine environment. It is characterized by a highly stratified sequence of sandstones, carbonates and shales (King et al., 1998, p. 318); the reservoir is generally thinly bedded and have highly variable heterogeneous reservoir and sedimentological characteristics (Dale, Lopes and Abilio 1990, p. 637). Pinda reservoirs are primarily dolomites with intercrystalline, moldic and vug porosity, and sandstones with intergranular porosity (Spaw and Koehler 1981, p. 996).

Ataliba Miguel

The average porosity for sandstone is typically 22% with permeabilities of 150 mD, whereas for carbonate facies is typically 10-20% and very relative low permeabilities of 10-20 mD (Dale, Lopes and Abilio 1990, p. 637).

3.3.2 Vermelha Reservoir

The Vermelha reservoir was deposited in a near-shore, shoreface and lagoonal settings of a coastal environment prevalent in Cenomanian time (King et al., 1998, p. 637).

The Vermelha reservoir consists of littoral sandstones with excellent reservoir properties deposited in a high-energy beach environment. These sandstones are of quartoze type, very fine to coarse grained, clean, consolidated to poorly consolidated, with intergranular porosity. The average porosity is about 25% to 35% and the permeability is 1000 mD. These sandstones grade laterally into well cemented dolomitic sandstones, with porosities of approximately 1% (Dale, Lopes and Abilio 1990, p. 636).

3.3.3 Likouala Reservoir

The Likouala reservoir was deposited in a littoral to coastal lagoon environment during the Cenomanian time. It consists of quartoze sandstones, white to reddish, with interbedded brown claystone (Lausaire, Makaya and Han 2009, p. 103).

3.4 Seal

Seals are commonly impermeable and porous. Shales and evaporites are the commonest types of seal. The sealing lithology across Block 0, are generally transgressive marine shales of Cenomanian-Eocene age (e.g. shales in the Bucomazi and Toca formations), and very rarely, thin supratidal dolomites and anhydrites (Schoellkopf and Patterson 2000, pp. 373-374).

Ataliba Miguel

3.5   Trap

Traps identified in the Congo Basin formed, as a result of graben development, resulting in both structural trap as shown in Fig 9, and stratigraphic trap as shown in Fig 10 (Berlinger et al., 2012, p.13).

The trap began to develop in the late Neocomian to Barremian, a period where rifting and subsidence developed in the central part of the Aptian salt basin (Brownfield and Charpentier 2006, cited by Edwards and Bignell, 1988b, and Teisserenc and Villemin 1990, p.33).

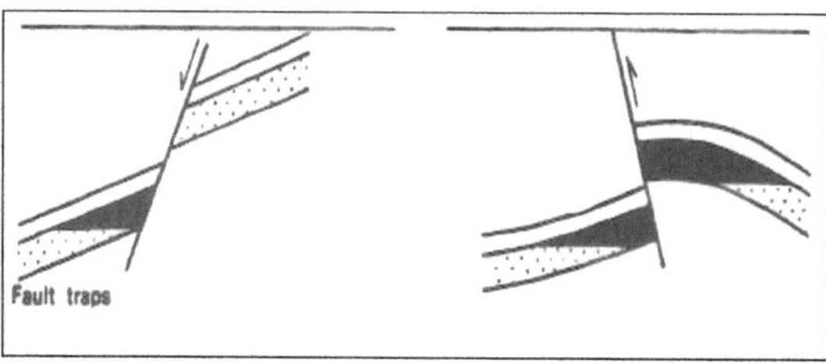

Figure 9 – Structural trap (adapted from Stoneley 1995)

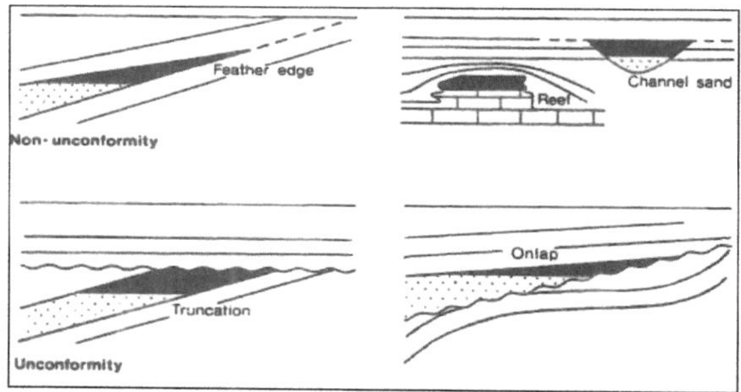

Figure 10 - Stratigraphic traps (adapted from Stoneley 1995)

Traps in the offshore Cabinda are formed in the Lucula and Toca (Fig 2) on upthrown tilted basement blocks, with Bucomazi shales and Loeme Salt acting as seals (Schoellkopf and Patterson 2000, p.23). Trap geometry is controlled by rollover anticlines and

Ataliba Miguel

associated faults. Traps associated with the Vermelha and Pinda Formations are rollover anticlines (Fig 11) in which an impermeable layer of marine shales overlies a permeable sandstone formation acting as well as a seal (Brownfield and Charpentier 2006, p. 33).

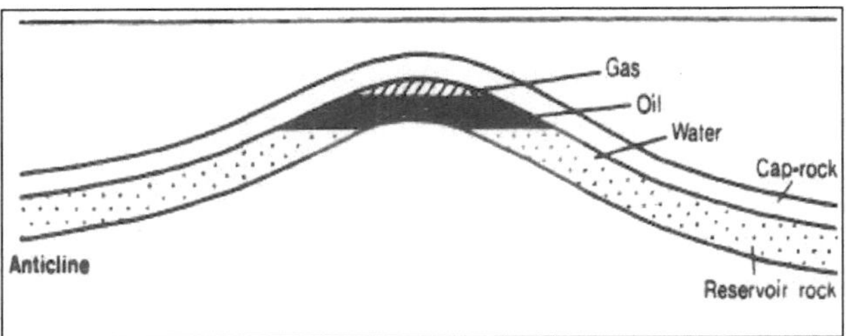

Figure 11 – Anticlinal Trap (adapted from Stoneley 1995)

4.   Prospectivity Impact on Block 0 – N'SANO Field

The N'SANO field prospectivity is analyzed as follows:

- Petroleum play
- Structural features
- Tectonic regime

Furthermore based on the petroleum system of Block 0 Area A, all the ingredients necessary for the generation, migration and accumulation are available.

The Bucomazi formation possesses very good source rocks, the porosity and permeability of the reservoir rock possesses very good characteristic associated with trap for hydrocarbon accumulations.

The impact on the prospectivity for the Bucomazi sourced hydrocarbons in Pinda reservoirs have been analyzed by Liro and Dawson (2000, pp. 85-86), and are summarized as follow:

- Pinda reservoir is a complex interplay of depositional and diagenetic processes, and the reservoir creation occurred in multiple episodes.

Ataliba Miguel

- Timing of trap development for Pinda reservoirs is partially structural due to halokinesis and partially stratigraphic due to secondary porosity development.
- The impermeable salt layer separates stratigraphically Bucomazi source rocks from Pinda reservoirs. Hydrocarbons accumulations in the Pinda reservoir was dependent both on the presence of salt evacuations or on the presence of growth faults; either case have facilitated the vertical migration from mature Bucomazi troughs

5.   Conclusion

The hydrocarbons from Block 0 area A are mainly sourced by the presalt Neocomian to Barremian Lacustrine shales of the Bucomazi formation.   The migration is mainly vertical (fault-related), the N'SANO field production comes from three non-communicating reservoir rocks; the Pinda, Vermelha and Likouala. The reservoirs are sealed by transgressive marine shales; and the trap is a combination of both structural and stratigraphic type. Table 2 summarizes the N'SANO play description. Figure 12 shows the timing events chart and Figure 13 illustrates a general conceptual model of a petroleum system.

Field	N'SANO
Source Rock	Bucomazi Formation
Migration	Vertical (fault related)
Reservoir rock	Pinda
	Vermelha
	Likouala
Seal	Marine Shales
Trap	Structural and Stratigraphic

Table 2– Block 0 N'SANO field petroleum system description.

   Ataliba Miguel

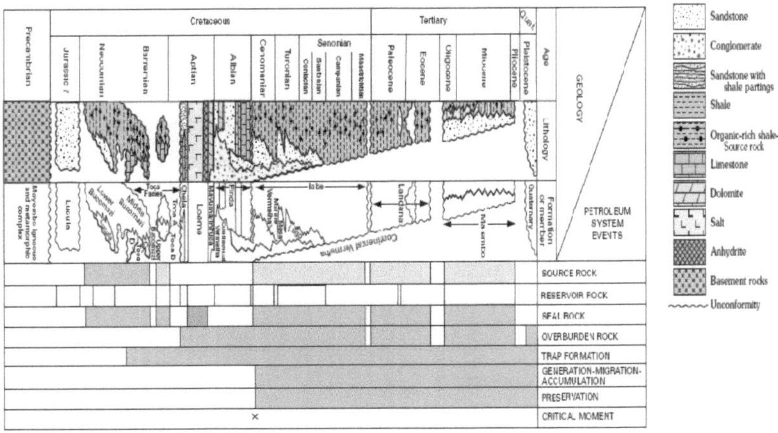

Figure 12 – Events chart for the Congo Delta Composite Total Petroleum System (Brownfield and Charpentier 2006)

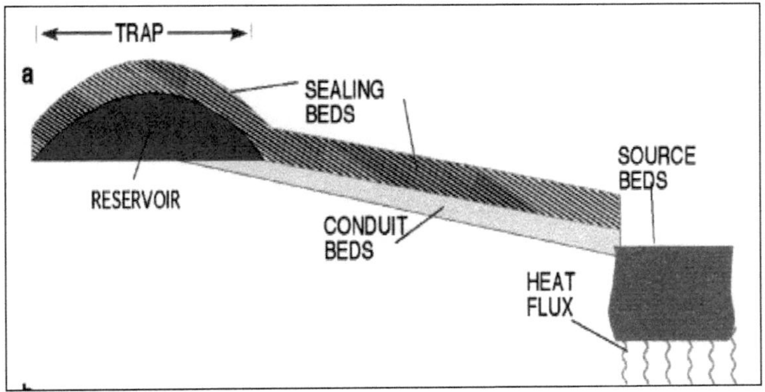

Figure 13 - General conceptual model of a Petroleum System (Aydin 2000)

Ataliba Miguel

6. References

Aydin, A. 2000. Fracture, faults and hydrocarbon entrapment, migration and flow. *Marine and Petroleum Geology,* 17 (7), pp. 797-814, Elsevier [Online]. Available at: http://www.sciencedirect.com/science/article/pii/S0264817200000 209#BIB22 (Accessed 12 April 2012).

Beglinger, E. S., Doust, H. and Cloetingh, S., 2011. Relating petroleum system and play development to basin evolution: West African South Atlantic basins. Marine and Petroleum Geology, 30(1), p. 1-25

Brice, S.E., et al., 1982. *Cole, G. A., et al., 2000:* Petroleum geochemical assessment of the Lower Congo Basin. In: M. R. Mello and B. J. Katz, eds. Petroleum systems of South Atlantic margins: AAPG Memoir 73, pp. 325–339.

Brownfield, M. E., and Charpentier, R. R., 2006, Geology and total petroleum systems of the West-Central Coastal Province (7203), West Africa. U.S. Geological Survey Bulletin 2207-B, p. 52

Burwood, R., 1999. *Cole, G. A., et al., 2000:* Petroleum geochemical assessment of the Lower Congo Basin. In: M. R. Mello and B. J. Katz, eds. Petroleum systems of South Atlantic margins: AAPG Memoir 73, pp. 325–339.

Burwood, R., et al,. 1995. *Cole, G. A., et al., 2000:* Petroleum geochemical assessment of the Lower Congo Basin. In: M. R. Mello and B. J. Katz, eds. *Petroleum systems of South Atlantic margins*: AAPG Memoir 73, pp. 325–339.

Cole, G. A., et al., 2000. Petroleum geochemical assessment of the Lower Congo Basin. In: M. R. Mello and B. J. Katz, eds. *Petroleum systems of South Atlantic margins*: AAPG Memoir 73, pp. 325–339.

Ataliba Miguel

England, W. A., Mann, A. L. and Mann, D. M., 1991. Migration from Source to Trap. In: Merril, R. K, ed. *Source and Migration Processes and Evaluation Techniques*: AAPG Special Volumes, pp. 23-46, [Online]. Available at: http://search.datapages.com/data/open/offer.do?target=%2Fspec pubs%2Fgeochem1%2Fdata%2Fa037%2Fa037%2F0001%2F0000 %2F0023.htm (Accessed on 12 March 2012)

Edwards and Bignell, 1988b. *Brownfield, M. E., and Charpentier, R. R., 2006.* Geology and total petroleum systems of the West-Central Coastal Province (7203), West Africa. U.S. Geological Survey Bulletin 2207-B, p. 52

Dale, C.T., Lopes, J.R. and Abilio, S., (1990). 'Takula Oil Field and the Greater Takula Area, Cabinda, Angola'. *The 22nd Annual OTC.* Houston, Texas, 7-10 May. Houston: Offshore Technology Conference, pp. 635-644.

King, G.R., et al., (1998) 'Reservoir Characterization, Geological Modelling, and Reservoir Simulation of the N'sano field, Upper Pinda Reservoir'. *Integrated Modelling for Asset Management.* Kuala Lumpur, Malaysia, 23-24 March. Kuala Lumpur: Society of Petroleum Engineers. pp. 317-326.

Lausaire, N., Makaya, N., and Han, C. H., 2009. Pre-Salt petroleum System of Vandji-Conkouati Structure (Lower Congo Basin), Republic of Congo. *Research of Journal of Applied Sciences*, 4 (3), pp. 101-107, Medwell Journals [Online]. Available at http://medwelljournals.com/abstract/?doi=rjasci.2009.101.107 (Accessed: 02 March 2012)

Liro, L. M., and W. C. Dawson., 2000. Reservoir systems of selected basins of the South Atlantic, *in* M. R. Mello and B. J. Katz,

     Ataliba Miguel

eds. *Petroleum systems of South Atlantic Margins*: AAPG Memoir 73, p. 77–92

Machado, V. (2007) Sand Provenance, Diagenesis and Hydrocarbon Charge History of the Kwanza Basin, Angola. Unpublished PhD thesis. University of Aberdeen.

Ping, L. J., et al., 2008. Petroleum geology and resources in West Africa: An overview. *Petroleum Exploration and Development*, 35(3), pp. 378-384, ScienceDirect[Online]. Available at: www.sciencedirect.com (Accessed: 04 Mach 2012).

Schoellkopf, N. B. and B. A. Patterson., 2000. Petroleum systems of offshore, Cabinda, Angola. In: M. R. Mello and B. J. Katz, eds. Petroleum systems of South Atlantic margins: AAPG Memoir 73, pp. 361–376.

Schlumberger (1993). *Going For the Play: Structural Interpretation in Offshore Congo.* [Online]. Available at: http://www.slb.com/resources/publications/oilfield_review/en/199 3/or1993_jan.aspx (Accessed: 29 March 2012)

Spaw, R. H. and Koehler, R. P., 1981. Geology and Reservoir Distribution, Pinda Formation, Offshore Zaire and Southern Offshore Cabinda: ABSTRACT. AAPG Bulletin, 30.

Teisserenc, P. and Villemin, J., 1990. *Brownfield, M. E., and Charpentier, R. R., 2006*. Geology and total petroleum systems of the West-Central Coastal Province (7203), West Africa. U.S. Geological Survey Bulletin 2207-B, p. 52

Bibliography:
Selley, R. C., 1998. Elements of Petroleum Geology, 2nd ed. London: Elsevier.

                                        Ataliba Miguel

Stoneley, R., 1995. An Introduction to Petroleum Exploration for Non-Geologists. New York: Oxford University Press.

Veeken, P. C. H., 2007. Handbook of Geophysical Exploration – Seismic Stratigraphy, Basin Analysis and Reservoir Characterisation, Volume 37. Knovel [Online]. Available at: http://www.knovel.com/web/portal/browse/display?_EXT_KNOVEL_DISPLAY_bookid=3192&VerticalIID=0 (Accessed 03 March 2012)

Ataliba Miguel